INTERNATIONAL TracTracTors

PHOTO ARCHIVE

INTERNATIONAL TracTracTors

PHOTO ARCHIVE

Photographs from the
McCormick-International Harvester Company Collection

Edited with introduction by
P.A. Letourneau

Iconografix
Photo Archive Series

Iconografix
PO Box 609
Osceola, Wisconsin 54020 USA

Library of Congress Card Number 96-76056

ISBN 1-882256-48-4

96 97 98 99 00 5 4 3 2 1

Cover design by Lou Gordon, Osceola, Wisconsin
Digital imaging by Pixelperfect, Madison, Wisconsin

Printed in the United States of America

Book trade distribution by Voyageur Press, Inc. (800) 888-9653

PREFACE

The histories of machines and mechanical gadgets are contained in the books, journals, correspondence and personal papers stored in libraries and archives throughout the world. Written in tens of languages, covering thousands of subjects, the stories are recorded in millions of words.

Words are powerful. Yet, the impact of a single image, a photograph or an illustration, often relates more than dozens of pages of text. Fortunately, many of the libraries and archives that house the words also preserve the images.

In the *Photo Archive Series*, Iconografix reproduces photographs and illustrations selected from public and private collections. The images are chosen to tell a story—to capture the character of their subject. Reproduced as found, they are accompanied by the captions made available by the archive.

The *Iconografix Photo Archive Series* is dedicated to young and old alike, the enthusiast, the collector and anyone who, like us, is fascinated by "things" mechanical.

The photographs and illustrations that appear in this book were made available by The State Historical Society of Wisconsin. The Society is the official repository for records of the International Harvester Company and its nineteenth-century predecessor, the McCormick Harvesting Machine Company.

The McCormick-International Harvester Company Collection contains nearly 4,000 cubic feet of family papers and business records, including technical publications, advertising literature, engineering and promotional photographs, posters, and films. Its cataloguing, preservation, and administration are funded through an endowment established in 1991 by Brooks McCormick.

Use of the collection is not generally restricted. However, due to its size and complexity, interested persons are encouraged to contact the Society in advance, at 816 State Street, Madison, Wisconsin 53706.

T-20 TracTracTor hauling logs for a mill in North Dana, Massachusetts. 1935.

INTRODUCTION

Greater traction and floatation, the principal advantages of track-laying machines, were recognized early in the history of the tractor. Patents for track-laying machines were recorded in Britain as early as 1770. The Minnis Steam Tractor, a track-layer built in Pennsylvania, was demonstrated in 1869. The first practical crawler tractors did not appear, however, before the early 1900s. Among the successful early manufacturers were Holt Manufacturing Company and C. L. Best Tractor Company. Holt, which built its first track-layers in 1904, and Best, which followed in 1913, eventually merged to form Caterpillar Tractor Company. Cleveland Tractor Company, another pioneer, introduced the first Cletrac machines in 1916.

The first track-equipped International Harvester tractors appeared around 1924, McCormick-Deering 10-20s and 15-30s fitted with half or full tracks manufactured by allied suppliers. Among the suppliers were Trackson, Moon Track, Hadfield-Penfield, and Mandt-Freil. Beginning in 1928, IHC built its own tracks and fitted them to 10-20 and 15-30 tractors. Designated as Model 10-20 and 15-30 Track Layers, these machines led to the design of the early TracTracTor.

The first TracTracTor was introduced in 1930. The Model 15 or T-15 was built in limited number, before it was succeeded by the T-20 in 1931. The T-20, with its 4-cylinder 3-3/4 x 5-inch gas engine and 3-speed transmission, developed a maximum 26.59 belt and 23.33 drawbar horsepower, and achieved maximum drawbar pull of 5,156 pounds in its Nebraska Tractor Test. The T-20 remained in production through 1939.

The TA-40 TracTracTor was introduced in 1932. It was succeeded in 1934 by the T-40. Early machines were equipped with a 3-5/8 x 4-1/2-inch 6-cylinder gas engine and 5-speed transmission. The TA-40 developed a maximum 46.48 belt and 41.78 drawbar horsepower, and achieved maximum drawbar pull of 9,399 pounds in its Nebraska Tractor Test. Later machines were equipped with a 3-3/4 x 4-1/2-inch 6-cylinder engine. The diesel TD-40, introduced in 1933, featured the same engine as fitted to the WD-40 wheel tractor. This 4-3/4 x 6-1/2-inch 4-cylinder diesel featured magneto, carburetor and spark plugs, as well as injection pump and injectors. The engine was started on gasoline and once up to speed operated on diesel fuel. The TD-40, equipped with 5-speed transmission, developed a maximum 53.46 brake and 48.53 drawbar horsepower, and achieved maximum drawbar pull of 10,487 pounds in its Nebraska Tractor Test. The T-40 and TD-40 remained in production through 1939.

The T-35 and TD-35 TracTracTors were introduced in 1937 and were built through 1939. Smaller versions of the T-40 and TD-40, both were fitted with 5-speed transmissions. The T-35 was equipped with a 3-5/8 x 4-1/2-inch 6-cylinder gas engine. It developed a maximum 44.44 brake and 36.58 drawbar horsepower, and achieved maximum drawbar pull of 8,053 pounds in its Nebraska Tractor Test. The TD-35 was equipped with a 4-1/2 x 6-1/2-inch 4-cylinder diesel engine that was started on gasoline and switched over to diesel fuel in the same manner as the TD-40.

In 1939, a new series of TracTracTors was introduced. Styled in a more streamlined fashion by the famed industrial designer Raymond Loewy, the tractors featured new engines, transmissions, and main frames, as well as numerous mechanical, electrical, and hydraulic improvements. The T-6 and diesel TD-6 were built from 1939 through 1956. The T-6, equipped with a 3-7/8 x 5-1/4-inch 4-cylinder gas engine and 5-speed transmission, developed maximum 36.96 brake horsepower, and achieved maximum drawbar pull of 7,653 pounds in its Nebraska Tractor Test. The TD-6, equipped with a 3-7/8 x 5-1/4-inch 4-cylinder diesel engine and 5-speed transmission, developed maximum 34.54 brake horsepower, and achieved maximum drawbar pull of 7,160 pounds in its Nebraska Tractor Test.

The T-9 and TD-9 were built from 1939 through 1956. The T-9, equipped with a 4-4/10 x 5-1/2-inch 4-cylinder gas engine and 5-speed transmission, developed maximum 46.03 brake horsepower and achieved maximum drawbar pull of 9,868 pounds in its Nebraska Tractor Test. The TD-9, equipped with a 4-4/10 x 5-1/2-inch 4-cylinder diesel engine and 5-speed transmission, developed maximum 43.93 brake horsepower and achieved maximum drawbar pull of 9,014 pounds in its Nebraska Tractor Test.

The T-14 was built from 1939 through 1946; the TD-14 from 1939 through 1948. The T-14 was equipped with a 4-3/4 x 6-1/2-inch 4-cylinder gas engine and 6-speed transmission. Fewer than 120 T-14s were built before production was suspended, and the tractor never participated in the Nebraska Tractor Test. The TD-14,

equipped with a 4-3/4 x 6-1/2-inch 4-cylinder diesel engine and 6-speed transmission, developed maximum 61.56 brake horsepower, and achieved maximum drawbar pull of 13,426 pounds in its Nebraska Tractor Test. The TD-14's successor, the TD-14A remained in production until 1955.

The TD-18 was built from 1939 through 1948. It featured a 6-speed transmission and 4-3/4 x 6-1/2-inch 6-cylinder diesel engine. The TD-18 developed a maximum 80.32 brake and 72.38 drawbar horsepower, and achieved maximum drawbar pull of 18,973 pounds in its Nebraska Tractor Test. The TD-18's successor, the TD-18A remained in production until 1955.

Production of the TD-24 began in 1947. The tractor tested at Nebraska weighed 40,595 pounds. Equipped with an 8-speed transmission and 5-3/4 x 7-inch 6-cylinder diesel, it developed a maximum 138.13 drawbar horsepower and achieved a maximum drawbar pull of 33,714 pounds. The TD-24 weighed nearly two tons more than a Caterpillar D-8. Its maximum drawbar pull exceeded that of the D-8 by more than 5,000 pounds.

Although most of the tractors in this later series remained in production well into the 1950s, the TracTracTor designation seems to have been dropped soon after the 1947 introduction of the TD-24. For this reason, *International TracTracTors Photo Archive* features no machines built after 1948.

McCormick-Deering 10-20 equipped with Hadfield-Penfield Steel Company *Alwatrac* full crawlers.

M-D 10-20 equipped with Moon Track semi-crawlers in a California orchard. 1924.

M-D 10-20 equipped with Moon Track semi-crawlers. Whittier, California, 1924.

M-D 10-20 with unidentified tracks working on road construction in Quebec, Canada. 1926.

Trackson-equipped McCormick-Deering 10-20.

M-D 10-20 working on road construction in Quebec, Canada.

M-D 10-20 equipped with Hadfield-Penfield *Alwatrac* full crawlers.

M-D 10-20 equipped with Mandt-Freil tracks and Monarch loader. California, 1927.

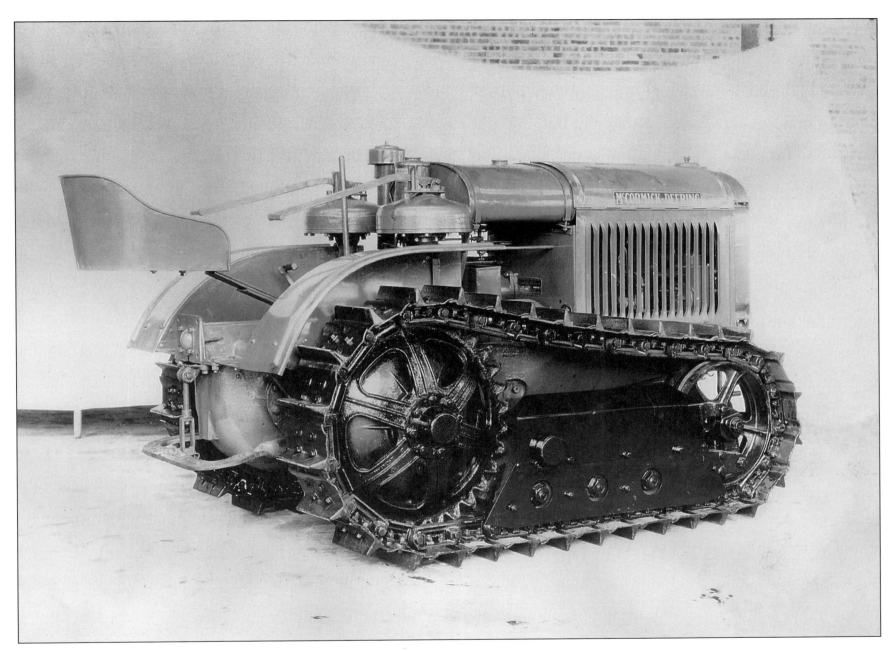

Early McCormick-Deering prototype TracTracTor based on a 10-20. 1928.

Model 15-30 Track Layer showing mechanism. June 1928.

Model 15-30 Track Layer. June 1928.

Two views of the TracTracTor Model 15 (T-15). June 1930.

Front view of the Model T-15. June 1930.

24

Model T-15 steering clutch assembly. Note pto shaft. June 1930.

Two views of the TracTracTor Model 20 (T-20). May 1930.

26

Front and side views of a T-20 fitted with full orchard fenders. August 1930.

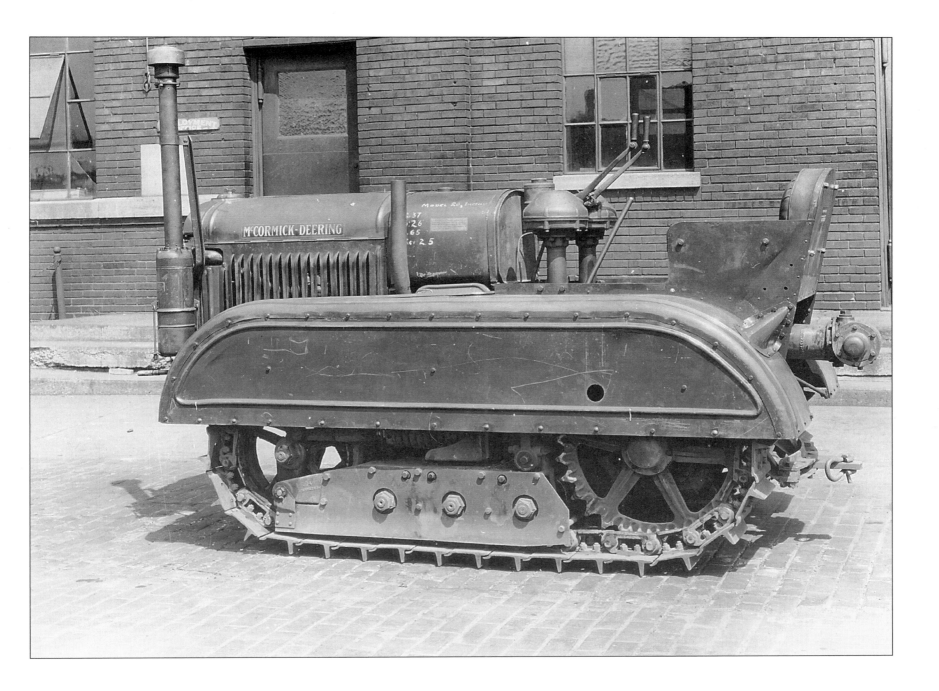

T-20 with rare side-mounted seat and controls.

T-20 with low seat attachment.

T-20 clearing Pennsylvania Railroad right-of-way.

T-20 with Bucyrus-Erie dozer. 1937.

T-20 orchard tractor pulling 7-foot disk in a California walnut grove. 1935.

T-20 pulling a sprayer through a 400 acre grape vineyard. Orange Grove, California, 1930.

T-20 switching railcars. Fort Williams, Ontario, 1937.

T-20 with reversed Whitehead & Kales crane. Pittsfield, Massachusetts, 1937.

Laying pipe with a T-20 equipped with sideboom. 1937.

T-20 working on a Pennsylvania Railroad electrification project near Baltimore, Maryland.

T-20 with McEwen winch operating in a Winfield, West Virginia forest.

T-20 and 8-foot McCormick-Deering combine in a rye field. North Carolina, 1935.

T-20 and Hardie sprayer in a grapefruit grove. Escondido, California, 1936.

T-20 hauling garbage. 1933.

Laying wire with a T-20 and cable plow. Louisville, Kentucky, 1937.

Illustration of a TracTracTor Model 40 (T-40). 1934.

Illustration of a diesel-powered TD-40. 1933.

Two views of a T-40 with factory cab. 1933.

Side and rear views of a TA-40. 1932.

T-40 with Schlusser-McLean dozer. 1933.

TD-40 pulling stumps. Hammon, Louisiana, 1938.

T-40 moves a carnival wagon. 1941.

A TD-40 and rock crusher ford a river near Elk Lake, Ontario. Note Koehring steam shovel. 1939.

T-40 with Willamette angle blade builds roads at a Civilian Conservation Corps camp near Deer Creek, Washington. 1933.

TD-40 with Sargent snow plow. Winnipeg, Manitoba, 1939.

T-40 plowing at the Salvation Army Farm, Wheaton, Illinois.

Two views of a T-40 equipped with an unidentifed angle dozer. 1934.

A new T-40 awaits delivery in front of the Rio Vista, California dealership. 1932.

A T-40 tows a Sikorsky airplane at the National Air Races, July 1933.

T-40 and an unidentified road grader.

T-40 pulling wagon loads of beets. Oxnard, California, 1933.

Unloading wagon loads of beets. Each wagon weighed 3,500 lb. Oxnard, California, 1933.

Illustrations of the TracTracTor Model 35 (T-35). 1937.

Illustrations of the TracTracTor Diesel Model 35 (TD-35). 1937.

Illustration of the rear view of the TracTracTor T-35. 1937.

Assembly workers installing the engine of a TD-35. 1937.

Factory test of a T-35 in an oil bath. 1937.

TD-35 widetread.

Loading a TD-35 on a rail flatcar. Hudson, Ontario, 1937.

TD-35 with Bucyrus-Erie dozer.

TD-35 pulling a spring-tooth harrow. Campbellville, Ontario, 1938.

TD-35 with orchard seat and fenders pulling a 9-3/4-foot covered dish harrow in a California orange grove.

T-35 with canopy pulling a subsoiler and chisel plow. Huntington Beach, California, 1938.

T-35 equipped with Ingersoll-Rand air compressor at work on the Sante Fe Railroad. 1940.

TD-35 pulls the Pan American Airways *Bermuda Clipper* from its hanger to the water. Baltimore, Maryland, 1938.

TD-35 hauling logs near Fort Frances, Ontario. 1937.

TD-35 with McCormick-Deering Model 41T combine. Farmall F-20 in background. Fisher, Minnesota, 1938.

TD-35 with McCormick-Deering 24-6 double disk drill. McMinnville, Oregon, 1936.

TD-35 switching railcars. Winnipeg, Manitoba, 1937.

T-35 moving an oil rig near Bradford, Pennsylvania.

Diesel-powered International TD-6 building fire guards near Goleta, California. 1941.

TD-6 with a 1-1/4-yard Hough front end loader. 1945.

Gasoline-powered T-6 at work on a Hyattsville, Maryland housing development. 1941.

TD-6 hauling logs at an Indiana mill. 1946.

T-6 with Bucyrus-Erie blade grading for a playground. Downingtown, Pennsylvania, 1949.

TD-6 with 1/2-yard front end loader removes dirt from the shaft of a lead mine. Joplin, Missouri, 1944.

TD-9 and Luther winch pulling the rods of an oil well. 1940.

Diesel-powered TD-9 with Bucyrus-Erie dozer. 1946.

TD-9 building fire trails near Santa Barbara, Claifornia. 1941.

TD-9 damming creek beds on a sheep ranch near Cowley, Wyoming.

TD-9 at work in a Kansas City quarry. 1944.

TD-9 with a 3/4-yard Bucyrus-Erie dozer shovel loading trucks at a Pittsburg, Kansas coal yard. 1944.

TD-9 at work in the Allegheny oil fields of southwestern New York. 1941.

A TD-14 at work in the Allegheny oil fields of southwestern New York. 1941.

Diesel-powered TD-14 with Isaacson *Trackdozer* and WO-14 winch skidding logs in an Oregon forest. 1941.

Pre-production TD-14 tractor. 1939.

TD-14 clearing land for a begonia, gardenia, and orchid farm in Mandeville Canyon, Los Angeles.

TD-14 with Bucyrus-Erie dozer clears land for development in Dunedin, Florida. 1941.

Harvesting salt with a TD-14 with Bucyrus-Erie blade. Note U2 power unit in background. Wendover, Utah, 1942.

TD-18 with Isaacson *Hydraulic Trackdozer*, Carco winch and arch builds access roads in a Washington forest.

TD-18 with Bucyrus-Erie dozer blade clearing stumps. 1940.

TD-18 with pipe boom. 1940.

TD-18 with Bucyrus-Erie dozer covers pipeline built to drain water from the bottom of a sand pit.

TD-18 with Bucyrus-Erie blade salvages refuse at Carnegie-Illinois Steel. Pittsburgh, Pennsylvania, 1944.

TD-18 with Isaacson *DIW* straight blade cable dozer, WO-18 winch hoist, and KA-Medum arch. Oregon, 1947.

TD-18 and LeTourneau scraper working on a 4-lane highway near St. Joseph, Missouri. 1942.

Two views of a TD-18 with cab hauling freight. 1939.

TD-18 with a 5-ton front and overshot scoop bucket loading sugar beets. Mullen, Nebraska, 1945.

TD-24 with Bucyrus-Erie angle dozer builds access roads in an Oregon forest.

TD-24 with Bucyrus-Erie dozer strips over-burden from a seam of coal.

TD-24 with Bucyrus-Erie scraper on a county fill project near Plymouth, Wisconsin.

TD-24 and Bucryus-Erie 250 scraper operating at an electric power plant. Essexville, Michigan, 1948.

TD-24 arching out a Douglas Fir measuring 32-feet long. Oregon, 1948.

TD-24 bulldozing coal in a Somerset, Massachusetts yard.

TD-24 with Bucyrus-Erie angle dozer and Isaacson rear winch builds roads in Montana's Lolo National Forest.

TD-24 and a 74-foot long gang of five Graham-Hoeme plows. Havre, Montana, 1948.

The Iconografix Photo Archive Series includes:

TRACTORS AND CONSTRUCTION EQUIPMENT

CASE TRACTORS 1912-1959 Photo Archive	ISBN 1-882256-32-8
CATERPILLAR MILITARY TRACTORS VOLUME 1 Photo Archive	ISBN 1-882256-16-6
CATERPILLAR MILITARY TRACTORS VOLUME 2 Photo Archive	ISBN 1-882256-17-4
CATERPILLAR SIXTY Photo Archive	ISBN 1-882256-05-0
CATERPILLAR THIRTY Photo Archive	ISBN 1-882256-04-2
CLETRAC AND OLIVER CRAWLERS Photo Archive	ISBN 1-882256-43-3
FARMALL F-SERIES Photo Archive	ISBN 1-882256-02-6
FARMALL MODEL H Photo Archive	ISBN 1-882256-03-4
FARMALL MODEL M Photo Archive	ISBN 1-882256-15-8
FARMALL REGULAR Photo Archive	ISBN 1-882256-14-X
FARMALL SUPER SERIES Photo Archive	ISBN 1-882256-49-2
FORDSON 1917-1928 Photo Archive	ISBN 1-882256-33-6
HART-PARR Photo Archive	ISBN 1-882256-08-5
HOLT TRACTORS Photo Archive	ISBN 1-882256-10-7
INTERNATIONAL TRACTRACTORS Photo Archive	ISBN 1-882256-48-4
JOHN DEERE MODEL A Photo Archive	ISBN 1-882256-12-3
JOHN DEERE MODEL B Photo Archive	ISBN 1-882256-01-8
JOHN DEERE MODEL D Photo Archive	ISBN 1-882256-00-X
JOHN DEERE 30 SERIES Photo Archive	ISBN 1-882256-13-1
MINNEAPOLIS-MOLINE U-SERIES Photo Archive	ISBN 1-882256-07-7
OLIVER TRACTORS Photo Archive	ISBN 1-882256-09-3
RUSSELL GRADERS Photo Archive	ISBN 1-882256-11-5
TWIN CITY TRACTOR Photo Archive	ISBN 1-882256-06-9

TRUCKS

DODGE TRUCKS 1929-1947 Photo Archive	ISBN 1-882256-36-0
DODGE TRUCKS 1948-1960 Photo Archive	ISBN 1-882256-37-9
MACK MODEL AB Photo Archive	ISBN 1-882256-18-2
MACK MODEL B 1953-1966 VOLUME 1 Photo Archive	ISBN 1-882256-19-0
MACK MODEL B 1953-1966 VOLUME 2 Photo Archive	ISBN 1-882256-34-4
MACK EB-EC-ED-EE-EF-EG & DE 1936-1951 Photo Archive	ISBN 1-882256-29-8
MACK EH-EJ-EM-EQ-ER-ES 1936-1950 Photo Archive	ISBN 1-882256-39-5
MACK FC, FCSW & NW1936-1947 Photo Archive	ISBN 1-882256-28-X
MACK FG-FH-FJ-FK-FN-FP-FT-FW 1937-1950 Photo Archive	ISBN 1-882256-35-2
MACK LF-LH-LJ-LM-LT 1940-1956 Photo Archive	ISBN 1-882256-38-7
STUDEBAKER TRUCKS 1927-1940 Photo Archive	ISBN 1-882256-40-9
STUDEBAKER TRUCKS 1941-1964 Photo Archive	ISBN 1-882256-41-7

The Iconografix Photo Archive Series is available from direct mail specialty book dealers and bookstores worldwide, or can be ordered from the publisher. For additional information or to add your name to our mailing list contact:

Iconografix
PO Box 609
Osceola, Wisconsin 54020 USA

Telephone: (715) 294-2792
(800) 289-3504 (USA)
Fax: (715) 294-3414

Book trade distribution by Voyageur Press, Inc., PO Box 338, Stillwater, Minnesota 55082 USA (800) 888-9653